MYSTERIES OF SPACE

INVESTIGATE!

WHY DOES THE MOON ORBIT EARTH?

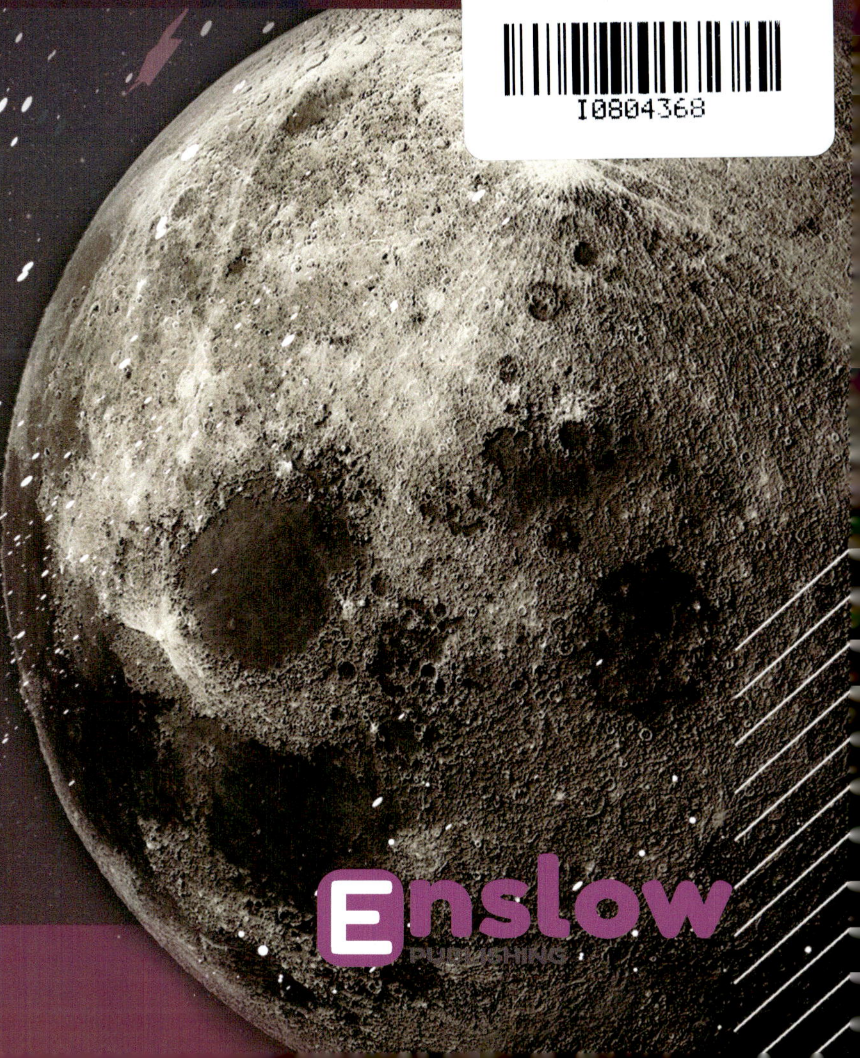

SOPHIE WASHBURNE

Enslow PUBLISHING

Please visit our website, www.enslow.com.
For a free color catalog of all our high-quality books, call toll free 1-800-398-2504 or fax 1-877-980-4454.

Cataloging-in-Publication Data
Names: Washburne, Sophie.
Title: Why Does the Moon Orbit Earth? / Sophie Washburne.
Description: New York : Enslow Publishing, 2023. | Series: Mysteries of Space | Includes glossary and index.
Identifiers: ISBN 9781978532991 (pbk.) | ISBN 9781978533011 (library bound) | ISBN 9781978533028 (ebook)
Subjects: LCSH: Earth (planet) –Juvenile literature. | Moon–Juvenile literature. | Natural satellite–Juvenile literature
Classification: LCC QB582 W38 2023 | DDC 523.3/3–dc23

Portions of this work were originally authored by Michael Portman and published as *Where Did the Moon Come From?*. All new material this edition authored by Sophie Washburne.

Published in 2023 by
Enslow Publishing
2544 Clinton Street
Buffalo, NY 14224

Designer: Tanya Dellaccio
Editor: Jennifer Lombardo

Photo credits: Cover (logo) AAVAA/Shutterstock.com; series background Molibdenis/Shutterstock.com; cover janez volmajer/Shutterstock.com; p. 5 (top) SmartS/Shutterstock.com; p. 5 (bottom) Josef_Svoboda/Shutterstock.com; p.6 courtesy of NASA; p. 7 NASA/JPL-Caltech/SwRI/MSSS; p. 9 VectorMine/Shutterstock.com; p. 11 (right) Nicku/Shutterstock.com; p. 11 (left) Michael Landrum/Shutterstock.com; p. 13 HelenField/Shutterstock.com; p. 14 Elena11/Shutterstock.com; p. 15 Kit Leong/Shutterstock.com; p. 16 https://upload.wikimedia.org/wikipedia/commons/6/66/Moon_rock_2%2C_JSC.jpg; p. 17 https://upload.wikimedia.org/wikipedia/commons/9/98/Aldrin_Apollo_11_original.jpg; p. 19 (bottom) Bencemor/Shutterstock.com; p. 19 (top) Dignity100/Shutterstock.com; p. 21 (bottom) Damsea/Shutterstock.com; p. 21 (top) Porcupen/Shutterstock.com; p. 23 (top) robert_s/Shutterstock.com; p. 23 (bottom) New Africa/Shutterstock.com; p. 25 Dima Zel/Shutterstock.com; p. 27 (top) Dotted Yeti/Shutterstock.com; p. 27 (bottom) Designua/Shutterstock.com; p. 29 https://upload.wikimedia.org/wikipedia/commons/4/4a/Artist%27s_concept_of_collision_at_HD_172555.jpg.

Printed in the United States of America

CPSIA compliance information: Batch #CWENS23: For further information contact Enslow Publishing at 1-800-398-2504.

CONTENTS

ANCIENT STORIES 4
IN GOOD COMPANY 6
AN IMPORTANT JOB 8
A CLOSER LOOK 10
THE YOUNG MOON 12
QUIET AND UNCHANGING 14
VISITING THE MOON 16
WHAT ROCKS CAN TELL US 18
WHERE DID IT COME FROM? 20
MADE AT THE SAME TIME 22
CAPTURED BY GRAVITY 24
A CRASH COURSE 26
STILL STUDYING 28
GLOSSARY 30
FOR MORE INFORMATION 31
INDEX 32

Words in the glossary appear in **bold** type the first time they are used in the text.

ANCIENT STORIES

For thousands of years, people have been telling stories about the moon. Some people say they can see a man's face in the moon. In many Asian countries, people say it's a rabbit instead of a man. Some ancient societies said the moon was controlled by a goddess.

Even though we know more about the moon today, people still tell stories about it that we can't prove. For example, some people say the full moon makes people act strange.

EXPLORE MORE!

THE WORD "LUNAR" IS USED WHEN TALKING ABOUT THINGS HAVING TO DO WITH THE MOON.

PEOPLE USED TO THINK WOLVES WERE MORE LIKELY TO HOWL AT THE MOON WHEN IT WAS FULL THAN AT ANY OTHER TIME. THIS HELPED CREATE THE **MYTH** OF THE WEREWOLF.

IN GOOD COMPANY

Our moon was the first one people knew about, but it isn't the only moon. In fact, there are more than 200 moons in our solar system! Some planets have a lot of moons. Others have none. Earth is the only planet we know of that has just one moon.

Not all moons are the same size. The biggest one we know of is Ganymede, one of Jupiter's moons. It's twice as big as our moon. It's even bigger than Mercury!

LIKE GANYMEDE, OUR MOON HAS A LOT OF **CRATERS.**

IN 2021, NASA TOOK THE FIRST UP-CLOSE PHOTOS OF GANYMEDE.

EXPLORE MORE!

OUR WORD "MOON" COMES FROM LATIN WORDS THAT MEAN "MEASURE" AND "MONTH." THIS IS BECAUSE PEOPLE USED TO KEEP TRACK OF THE MONTHS BY LOOKING AT THE MOON'S PHASES.

AN IMPORTANT JOB

The moon helps make Earth a good place to live. Because of **gravity**, everything in space is in a constant game of tug-of-war. The moon's gravity pulls on Earth, and Earth's gravity pulls on the moon.

Because Earth isn't perfectly round, it wobbles as it spins, or rotates. The moon's gravity slows Earth's wobble. This has allowed Earth to have comfortable and steady weather. Without the moon, Earth would spin much faster, and the days would be much shorter.

EXPLORE MORE!

THE MOON'S GRAVITY AFFECTS THE TIDES IN THE OCEAN. THE PLANTS AND ANIMALS THAT LIVE IN THE WATER WOULD DIE WITHOUT THESE TIDES.

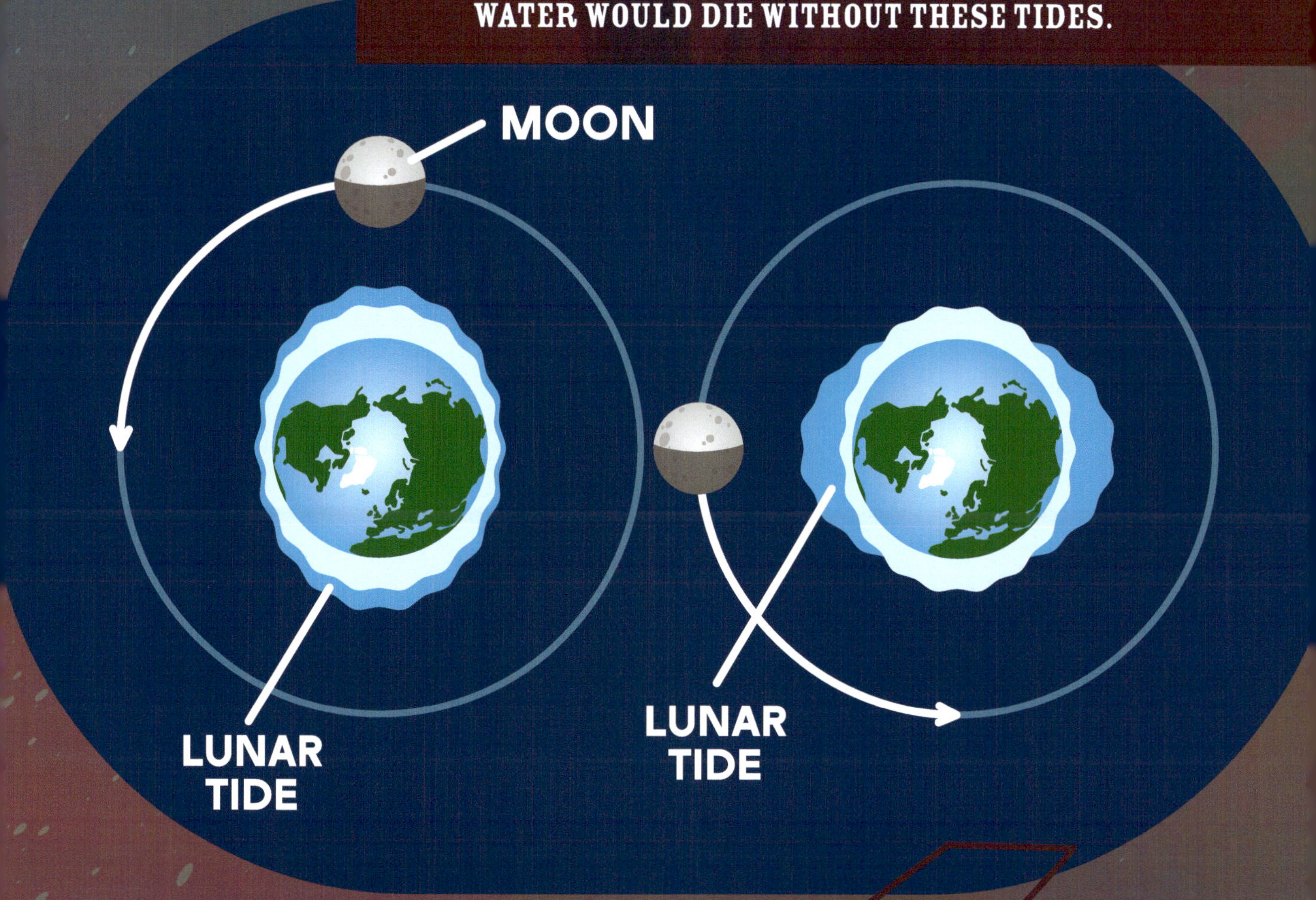

THIS PICTURE SHOWS HOW THE MOON AFFECTS THE TIDES.

A CLOSER LOOK

When we look at the moon, we see areas of light and dark. Long ago, some people believed the light areas were solid land and the dark areas were oceans. This is why the dark areas are called *maria*. This is the Latin word for "seas."

The Italian **astronomer** Galileo was the first person to use a **telescope** to study the moon. In 1609, Galileo noted that the moon isn't smooth. Instead, it's covered with mountains and valleys, much like Earth is.

EXPLORE MORE!

THE FIRST PEOPLE TO VISIT THE MOON LANDED IN A SPOT CALLED THE SEA OF TRANQUILITY. "TRANQUILITY" MEANS "CALMNESS." IT WAS NAMED IN THE 1600s BY ASTRONOMER GIAMBATTISTA RICCIOLI.

YOU DON'T NEED A TELESCOPE TO SEE THE SEA OF TRANQUILITY.

GALILEO

THE YOUNG MOON

The lunar maria aren't real seas today, but billions of years ago, they were seas of hot, liquid rock. Liquid rock underground is called magma. When the moon was young, much of it was made of this material. When magma is pushed to the surface, it's called lava. The moon's lava cooled quickly to form flat, dark areas of rock.

For more than a billion years, lava continued to pour out of lunar **volcanoes**. This created the mountains and valleys Galileo saw. The lighter parts of the moon are called the lunar highlands. They're rocky and covered with craters.

EXPLORE MORE!

THE MOON TAKES THE SAME AMOUNT OF TIME TO ROTATE AS IT DOES TO **ORBIT** EARTH ONCE. FOR THIS REASON, WE ALWAYS SEE THE SAME SIDE OF THE MOON.

THERE ARE THOUSANDS OF CRATERS ON THE MOON'S SURFACE.

QUIET AND UNCHANGING

The craters on the moon formed when rocky objects called asteroids hit the moon's surface. We don't have a lot of craters on Earth because most asteroids burn up in the planet's thick **atmosphere**. The moon has a very thin atmosphere, so it also has no weather or air to breathe.

Without wind or liquid water, the moon's surface changes very, very slowly. Dust on the moon can sit in the same place for hundreds of years. Nothing can grow or live there.

THE FOOTPRINTS ASTRONAUTS HAVE LEFT ON THE MOON ARE LIKELY STILL THERE.

EXPLORE MORE!

IN 2022, SCIENTISTS GREW PLANTS IN LUNAR SOIL FOR THE FIRST TIME. HOWEVER, THESE PLANTS COULDN'T HAVE GROWN WITHOUT THE AIR AND WATER WE HAVE ON EARTH.

LARGER OBJECTS MAKE LARGER CRATERS, LIKE THIS CRATER IN ARIZONA.

VISITING THE MOON

Between 1969 and 1972, 12 astronauts walked on the moon as part of space missions. When they came back to Earth, they brought back samples for scientists to study. These samples included about 2,200 moon rocks as well as some of the moon's dirt.

These samples were important clues in learning how the moon was formed. Scientists didn't know exactly what the moon was made of until the astronauts brought samples back. The samples showed that the moon has rocks similar to those on Earth.

ROCKS FROM THE MOON DON'T LOOK MUCH DIFFERENT FROM EARTH'S ROCKS.

BUZZ ALDRIN

EXPLORE MORE!

THE FIRST TWO PEOPLE TO WALK ON THE MOON WERE AMERICAN ASTRONAUTS NEIL ARMSTRONG AND EDWIN "BUZZ" ALDRIN. THEIR SPACECRAFT LANDED ON THE SEA OF TRANQUILITY ON JULY 20, 1969.

WHAT ROCKS CAN TELL US

Volcanoes created most of the moon's rocks. We know this because the rocks the astronauts brought back are igneous. This is the word for solid rocks made when liquid rock cools.

The lunar maria are made of a type of igneous rock called basalt. The lunar highlands are mostly made of a type of igneous rock called anorthosite. The dusty dirt on the moon's surface is called regolith. It's created when asteroids crash into the moon and break up the rocks.

EXPLORE MORE!

SCIENTISTS BELIEVE THE MOON HAS A VERY SMALL CORE, OR CENTER, UNDER ITS ROCKY OUTER SHELL. THE CORE IS LIKELY MADE MOSTLY OF IRON, JUST LIKE EARTH'S CORE.

BASALT AND ANORTHOSITE ARE FOUND ON EARTH TOO.

ANORTHOSITE

BASALT

WHERE DID IT COME FROM?

People have been trying for hundreds of years to figure out where our moon came from. In the 1800s, some scientists believed Earth had once rotated, or spun, so fast that a large chunk broke off and became the moon. However, there's no reason to believe Earth ever rotated that fast. Also, other planets spin faster and don't lose parts of themselves.

Today, there are three main theories, or ideas, about how the moon was created. They're called the co-formation theory, the capture theory, and the giant **impact** theory.

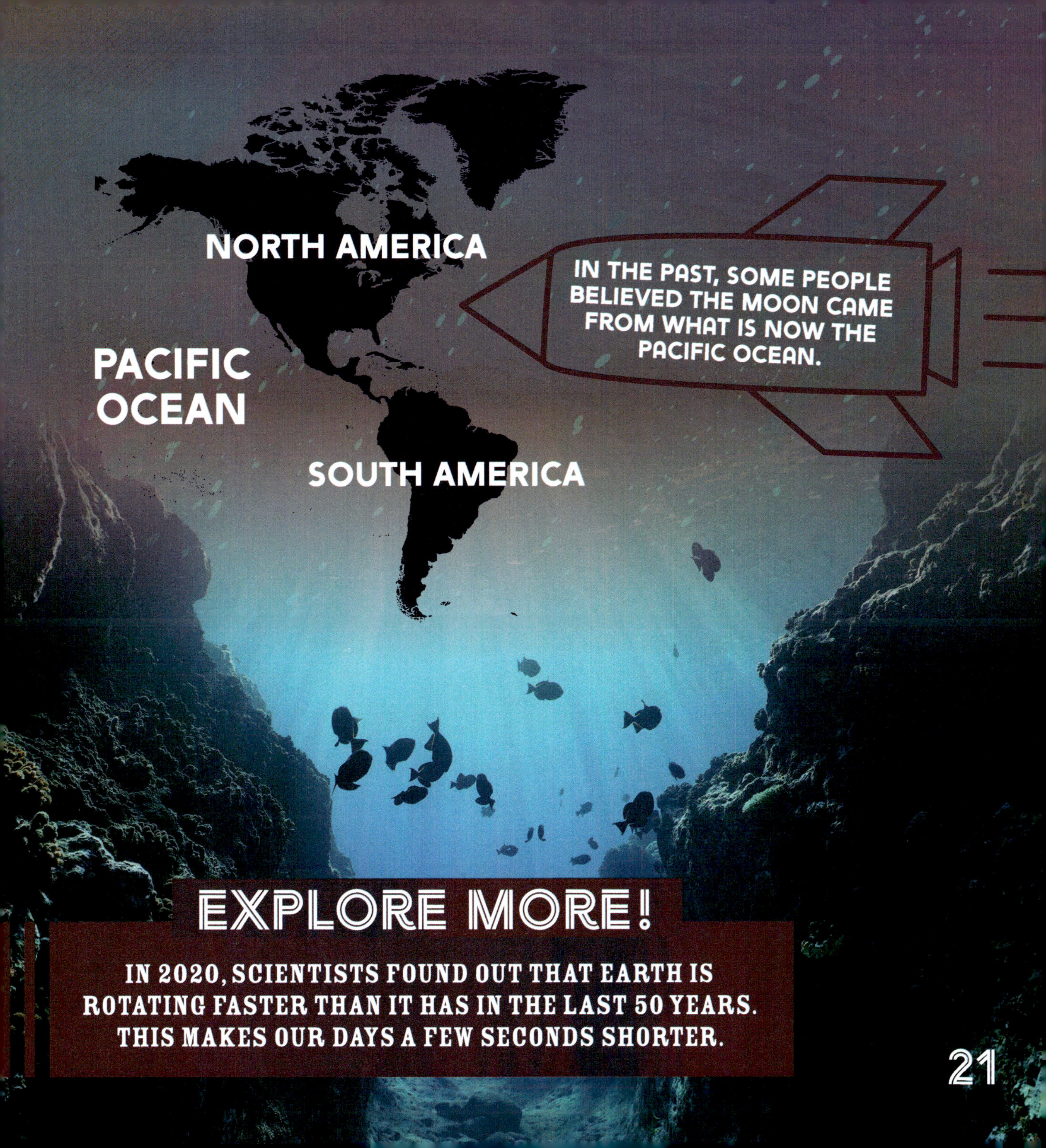

IN THE PAST, SOME PEOPLE BELIEVED THE MOON CAME FROM WHAT IS NOW THE PACIFIC OCEAN.

EXPLORE MORE!

IN 2020, SCIENTISTS FOUND OUT THAT EARTH IS ROTATING FASTER THAN IT HAS IN THE LAST 50 YEARS. THIS MAKES OUR DAYS A FEW SECONDS SHORTER.

MADE AT THE SAME TIME

The co-formation theory says that the moon and Earth formed near each other at the same time from the same matter. Scientists believe Earth formed by accretion. This means dust, rock, and gas were pulled by gravity into the shape of a ball.

Some people believe that some of this matter became Earth and some became the moon. However, the problem with the co-formation theory is that the moon and Earth aren't made of exactly the same material.

EXPLORE MORE!

SCIENTISTS HAVE FOUND THAT THE MOON'S SURFACE IS VERY SIMILAR TO EARTH'S. HOWEVER, IT GETS LESS SIMILAR CLOSER TO THE CORE.

ACCRETION BRINGS SMALL PIECES TOGETHER TO MAKE SOMETHING BIG, JUST LIKE MAKING A SNOWBALL.

CAPTURED BY GRAVITY

The capture theory says the moon formed somewhere else in the solar system. At some point, it traveled close enough to Earth to get pulled into its orbit by the planet's gravity.

This theory explains why the moon and Earth aren't exactly the same. However, it has other problems that make it an unlikely explanation. One problem is that Earth's gravity isn't strong enough to capture something the size of the moon slowly. The moon would have crashed into Earth instead.

IT'S NOT LIKELY THAT THE MOON APPEARED IN EARTH'S ORBIT FULLY FORMED.

EXPLORE MORE!

LARGER PLANETS, SUCH AS JUPITER AND SATURN, ARE MORE LIKELY TO HAVE CAPTURED THEIR MOONS. THIS IS BECAUSE LARGER PLANETS HAVE MORE GRAVITY.

A CRASH COURSE

The giant impact theory is our best guess about where the moon came from. Scientists came up with it in the mid-1970s. This idea says that a Mars-sized planet slammed into Earth while Earth was still forming. This planet was moving slowly enough that it didn't totally destroy Earth. Instead, only some parts of Earth were destroyed.

After the crash, magma and large pieces of both planets flew into space. The **debris** orbited Earth in a clump and finally formed a ball of magma. When it cooled, it was the moon we see today.

BEFORE THE MOON FORMED, THE DEBRIS MAY HAVE GIVEN EARTH RINGS LIKE SATURN'S.

EXPLORE MORE!

SCIENTISTS GAVE THE PLANET THAT MAY HAVE SMASHED INTO EARTH THE NAME "THEIA." THIS IS THE NAME OF THE MOTHER OF A GODDESS OF THE MOON IN GREEK MYTHS.

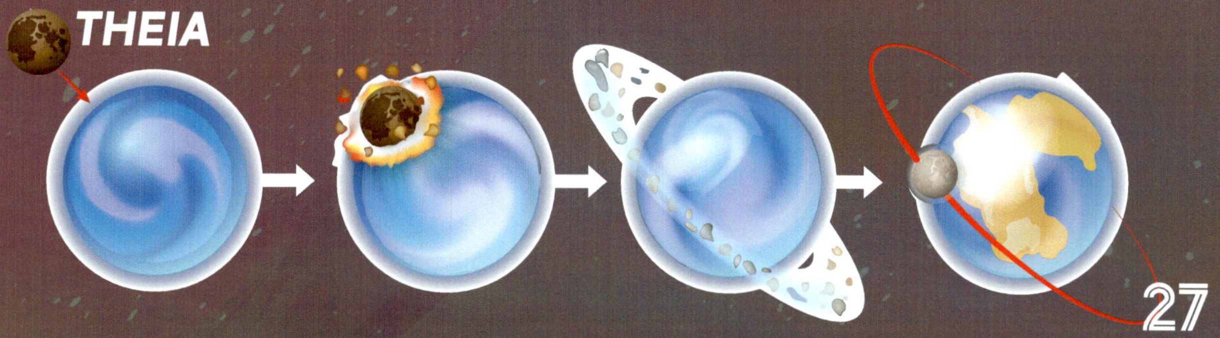

STILL STUDYING

The giant impact theory isn't perfect, but it's the theory we have the most **evidence** for. If a crash did happen, both Earth and the moon should have pieces of Theia in them.

For a long time, scientists couldn't find any evidence of Theia on either the moon or Earth. However, in 2020, they compared the **elements** in moon and Earth rocks and found differences that can be explained by the giant impact theory. What else will they find in the future?

THIS PICTURE SHOWS AN ARTIST'S IDEA OF WHAT THE IMPACT MAY HAVE LOOKED LIKE.

EXPLORE MORE!

THE GIANT IMPACT THEORY ALSO EXPLAINS WHY THE MOON IS SO BIG COMPARED TO THE EARTH. IT'S BECAUSE IT WAS FORMED FROM AN UNUSUALLY LARGE AMOUNT OF DEBRIS.

GLOSSARY

astronomer: A person who studies stars, planets, and other heavenly bodies.

atmosphere: The mixture of gases that surround a planet.

crater: A hole formed by an impact.

debris: The remains of something that has been broken.

element: A basic matter made of atoms only of one kind.

evidence: Something that helps show or disprove the truth of something.

gravity: The force that pulls objects toward the center of a planet, star, or moon.

impact: The action of one thing hitting another.

myth: A story that tries to explain mysterious events.

orbit: To travel in a circle or oval around something, or the path used to make that trip.

telescope: A tool that makes faraway objects look bigger and closer.

volcano: An opening in the ground from which melted rock and steam escape.

FOR MORE INFORMATION

BOOKS

Huddleston, Emma, and Larry Esposito. *Explore the Moon*. Minneapolis, MN: Abdo Publishing, 2022.

Ringstad, Arnold. *The Moon*. Mankato, MN: Child's World, 2021.

Rocco, John. *How We Got to the Moon: The People, Technology, and Daring Feats of Science Behind Humanity's Greatest Adventure*. New York, NY: Crown Books for Young Readers, 2020.

WEBSITES

NASA Space Place

spaceplace.nasa.gov/search/moon

Learn more about the moon and do some fun experiments at home.

National Geographic Kids: "Facts About the Moon!"

www.natgeokids.com/uk/discover/science/space/facts-about-the-moon

Read 10 fun facts about the moon.

Publisher's note to educators and parents: Our editors have carefully reviewed these websites to ensure that they are suitable for students. Many websites change frequently, however, and we cannot guarantee that a site's future contents will continue to meet our high standards of quality and educational value. Be advised that students should be closely supervised whenever they access the internet.

INDEX

accretion, 22, 23
Aldrin, Edwin "Buzz," 17
anorthosite, 18, 19
Armstrong, Neil, 17
asteroids, 14, 18
astronauts, 14, 16, 17, 18
basalt, 18, 19
capture theory, 20, 24
co-formation theory, 20, 22
craters, 6, 12, 13, 14, 15
Galileo, 10, 11
Ganymede, 6, 7
giant impact theory, 20, 26, 27, 28, 29
gravity, 8, 9, 22, 24, 25
highlands, 12, 18
Jupiter, 6, 25
magma, 12, 26
maria, 10, 12, 18
Mars, 26
Mercury, 6
phases, 5, 7
regolith, 18
Riccioli, Giambattista, 10
Saturn, 25, 27
Sea of Tranquility, 10, 11, 17
Theia, 27, 28
tides, 9
volcanoes, 12, 18
weather, 8, 14